'JUN 0 8

Graphic Organizers in Science™

Learning About Rocks, Weathering, and Erosion with Graphic Organizers

Diana Estigarribia

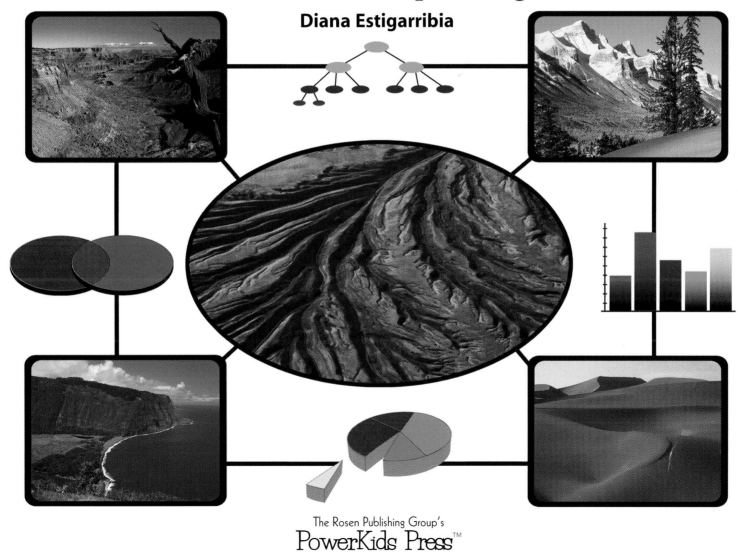

The Rosen Publishing Group's
PowerKids Press™
New York

For my parents, Concepción and Augusto, with love

Published in 2005 by The Rosen Publishing Group, Inc.
29 East 21st Street, New York, NY 10010

First Edition

Editor: Natashya Wilson
Book Design: Mike Donnellan

Photo Credits: Cover (center) © Annie Griffiths Belt/CORBIS; cover (top left) © CORBIS; cover (top right) © Eyewire; cover (bottom left and right) © DigitalVision; p. 4 © Danny Lehman/CORBIS; pp. 7 (top, third from top), 8 © Maurice Nimmo; Frank Lane Picture Agency/CORBIS; p. 7 (second from top) © Lester V. Bergman/CORBIS; p. 7 (second from bottom) © Anthony Bannister; Gallo Images/CORBIS; p. 7 (bottom) © Charles O'Rear/CORBIS; p. 12 (right) © Charles Mauzy/CORBIS; p. 12 (left) © W. Cody/CORBIS; p. 15 (bottom left) © Royalty Free/CORBIS.

Estigarribia, Diana
    Learning about rocks, weathering, and erosion with graphic organizers / Diana Estigarribia.
        p. cm. — (Graphic organizers in science)
    Includes bibliographical references and index.
    Summary: This book describes how scientists learn about the earth by studying different kinds of rocks and how they weather and erode.
    ISBN 1-4042-2806-3 (lib.) — ISBN 1-4042-5042-5 (pbk.)
    1. Weathering—Juvenile literature   2. Rocks—Juvenile literature   3. Geological time—Juvenile literature   [1. Weathering
2. Rocks   3. Geological time]   I. Title   II. Series
    QE570.E78 2005   2003-020124
    551.3—dc21

Manufactured in the United States of America

Contents

The Colorado River formed the Grand Canyon, in Arizona, over millions of years. The water is still carrying away rock bit by bit. This makes the huge canyon deeper.

Venn Diagram: Weathering and Erosion

Weathering

Erosion

- Softens rock
- Breaks down rock
- Creates soil, clay, sand, and gravel
- Caused by living things
- Caused by freezing and thawing

- Caused by water, ice, and wind
- Shapes land
- Creates new landforms

- Moves rock particles
- Helps weathering to continue
- Caused by gravity

Shaping Earth

The forces of weathering and erosion have shaped Earth's surface for billions of years. These forces are still changing the planet today. Weathering is the breakdown of rocks on the surface of Earth by water, wind, and ice. All the soil, clay, mud, gravel, and sand on Earth come from weathered rock.

Once rock is weathered, it is moved by the force called erosion. **Gravity** causes erosion because it pulls things downward. Rockslides are a form of erosion caused by gravity. Water, wind, and ice help gravity to erode rock. Once erosion carries away weathered rock, the rock left behind begins to weather, starting the process once again.

Graphic organizers are study tools that help to organize facts. In this book, graphic organizers are used to show how rock forms, breaks down, moves, and re-forms all over Earth.

A Venn diagram is a graphic organizer that shows how two things are alike and how they are different. Here the outside of each circle shows the different features of weathering and erosion. The features that both processes share go in the middle, where the circles overlap.

Minerals

Almost all rocks are made of one or more minerals. Minerals are made of solid, nonliving matter that occurs in nature. A mineral can be made of just one **element**. Elements are the basic matter that forms all things. Silver is a mineral made of just one element. Most minerals are made of two or more elements. When elements join, they form a **compound**. Quartz is a compound mineral. Minerals are **crystalline**. This means that the tiny **particles** that create a mineral join in a repeating pattern. When enough particles join, they form a mineral crystal.

Scientists who study minerals are called mineralogists. They have found about 2,000 different types of minerals. Mineralogists look at a mineral's hardness, color, and shine. They also look at how see-through a mineral is, and at its **chemical** features.

Charts are used to make facts easy to look up. This chart summarizes the features of the 10 common minerals used by German mineralogist Friedrich Mohs to create Mohs' scale of mineral hardness. Mohs rated the minerals from 1, the softest, to 10, the hardest. Harder minerals scratch softer minerals. All minerals on Earth can be rated on Mohs' scale.

Chart: Features of Common Minerals

Mineral	Mohs' Scale Hardness Rating	Description
Talc	1	Softest mineral, crumbles easily. Comes in white, green, brown, and gray. All other minerals can scratch talc.
Gypsum	2	Soft. Comes in clear and colors such as white, gray, orange, pink, and yellow. Thin crystals are bendable.
Calcite	3	Comes in white, yellow, or clear. Found in limestone and marble. Can be scratched by a penny.
Fluorite	4	Comes in all colors, including multicolored. Can be see-through. Can be scratched by a knife or a nail.
Apatite	5	Comes in clear and many colors, including white, green, red, and brown. Can be scratched by glass.
Feldspar	6	Comes in pink, white, and green. Commonly found in granite. Scratches glass.
Quartz	7	Comes in clear and colors such as white, purple, and brown. Scratches steel.
Topaz	8	Comes in clear and many colors, including yellow, orange, and brown. Scratches quartz.
Corundum	9	Second-hardest mineral. Comes in many colors, such as blue, red, and green. Sapphires and rubies are corundum.
Diamond	10	Hardest matter on Earth. Comes in clear or white. May have other colors in it, such as yellow. Scratches everything.

Cycle: The Rock Cycle

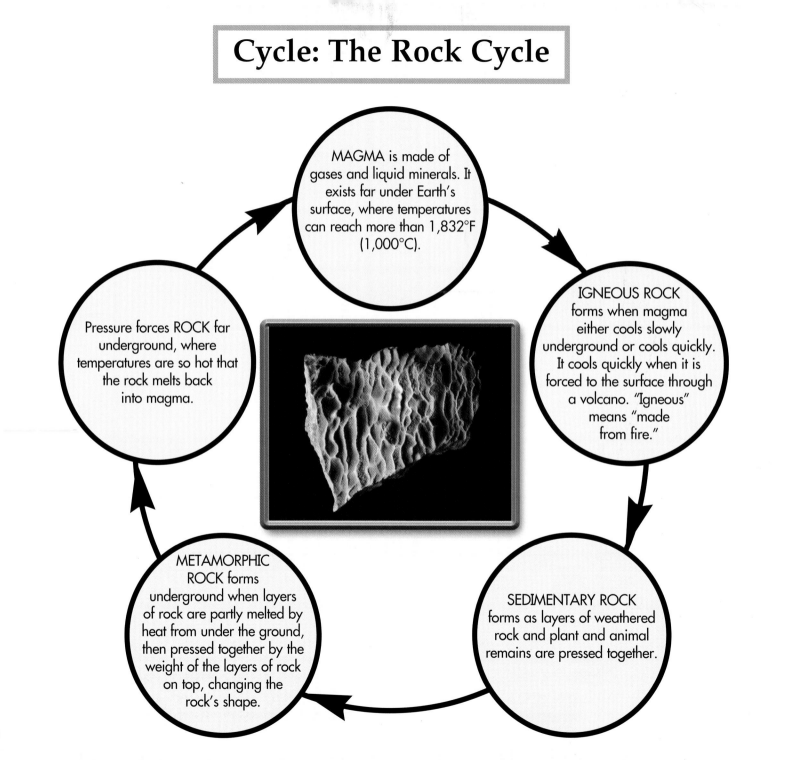

MAGMA is made of gases and liquid minerals. It exists far under Earth's surface, where temperatures can reach more than 1,832°F (1,000°C).

IGNEOUS ROCK forms when magma either cools slowly underground or cools quickly. It cools quickly when it is forced to the surface through a volcano. "Igneous" means "made from fire."

Pressure forces ROCK far underground, where temperatures are so hot that the rock melts back into magma.

SEDIMENTARY ROCK forms as layers of weathered rock and plant and animal remains are pressed together.

METAMORPHIC ROCK forms underground when layers of rock are partly melted by heat from under the ground, then pressed together by the weight of the layers of rock on top, changing the rock's shape.

Rocks and the Rock Cycle

Rocks are solid masses of matter that form naturally when minerals harden. Rocks are sorted into three groups. **Igneous rock** forms when hot, liquid **magma** from under the ground cools and becomes solid. Granite is an igneous rock formed by magma cooling very slowly underground. Obsidian and pumice are igneous rocks that form when magma shoots to Earth's surface through a volcano and cools quickly. **Sedimentary rock** forms over many thousands of years as layers of **sediment** press together and harden. Sediment is made up of tiny pieces of rock, such as mud or sand, and remains of dead plants and animals. Limestone is a sedimentary rock. **Metamorphic rock** is solid rock that has been changed by heat and pressure. Marble is a metamorphic rock formed from limestone. All rock changes from one type into another in a never-ending process called the rock cycle.

A cycle shows events that happen over and over again. Each arrow points to the next step. This cycle shows one path of possible steps in the rock cycle. All the rock on Earth changes from one type into another over millions of years. Center: *This is limestone.*

Chemical Weathering

All over Earth, rocks are broken down by the natural process of weathering. There are two types of weathering, called chemical weathering and physical weathering. In chemical weathering, water, air, and **acids** cause chemical changes in the minerals of the rocks. A chemical change happens when elements **dissolve** or re-form as new compounds. Water can dissolve minerals in rock. It can also combine with minerals to form new, softer minerals, weakening the rock. Clay is formed when water combines with minerals in rocks. Rain speeds up weathering by joining with gases in the air to form acids. An acid is a liquid that breaks down matter. Some living things also produce acids that break down rock. Air chemically weathers rock by oxidation. The oxygen in air joins with minerals in rock to form new compounds, making the rock crumble. Rust is an example of oxidation.

A classifying web can be used to sort a large group into smaller groups or to break down a subject into parts. This classifying web breaks down the subject, chemical weathering, into the three things that cause it to happen and their effects on rock.

Classifying Web: Chemical Weathering

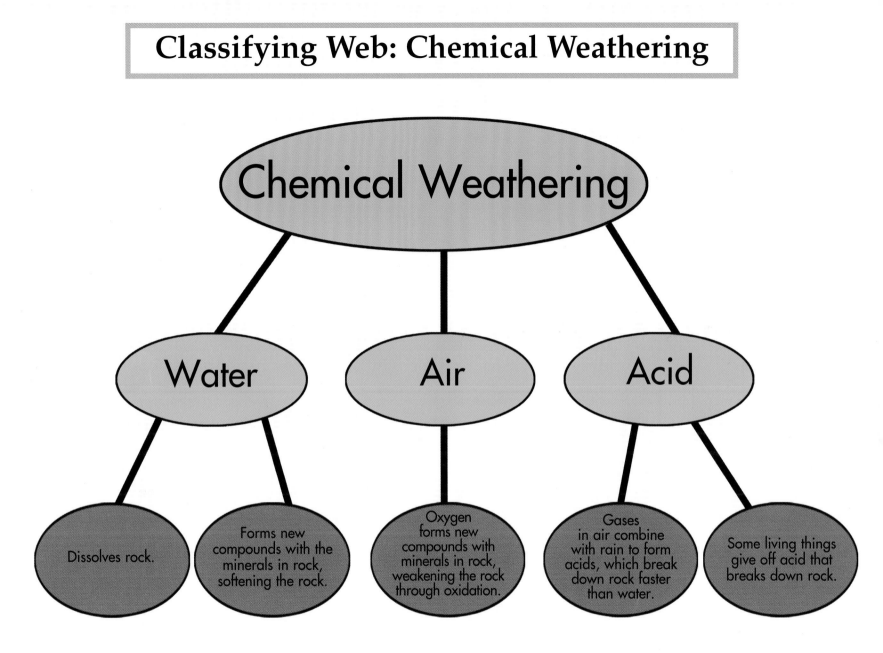

Chemical Weathering

Water

Air

Acid

Dissolves rock.

Forms new compounds with the minerals in rock, softening the rock.

Oxygen forms new compounds with minerals in rock, weakening the rock through oxidation.

Gases in air combine with rain to form acids, which break down rock faster than water.

Some living things give off acid that breaks down rock.

Compare/Contrast Chart: Chemical and Physical Weathering

	Chemical Weathering	Physical Weathering
Water	• Dissolves minerals in rock. • Combines with minerals to form new compounds, weakening rock. • Creates acid that breaks down rock.	• Freezes in the cracks and pores of rock, creating bigger cracks until the rock breaks apart.
Air	• Combines with minerals to form new compounds, weakening rock. • Pollution in air can combine with rain to make acids that break down rock.	• Wind rubs the particles it carries against rock, wearing away the rock.
Living Things	• Give off acids that break down rock.	• Plant roots crack rock. • Animals dig through the ground and break apart rock.

Physical Weathering

Physical weathering breaks down rock without changing it chemically. This happens as rocks crack and break apart into smaller pieces. Water is a large part of physical weathering. If water enters tiny cracks and holes in rocks and then freezes into ice, the ice takes up more space. The ice pushes against the rock and forces it to crack. Scientists believe that large changes in **temperature** may also crack and break down rock. This process takes thousands of years. Living things also cause physical weathering. Plant roots can split rock. Animals such as worms and gophers dig holes in the ground and move rocks around. The rocks rub against other rocks, causing more cracks and breaks. Wind physically weathers rock by blowing rock particles against it. The particles hit the rock and wear it away. This process is called **abrasion**.

A compare/contrast chart compares features of different things. This chart compares how water, air, and living things affect rock chemically and how they affect it physically. Top Right: Plant roots crack and break up rock. Top Left: Water and rock can form clay.

Erosion

Once rock is weathered into sediment, the sediment is moved to new places by erosion. Erosion works through gravity, wind, water, and ice. Gravity can cause sediment to move slowly down a hill or quickly down a cliff. This movement of sediment by gravity is called **mass wasting**. Plants slow down mass wasting. Their roots keep soil in place longer. Erosion happens much faster on land that has been cleared of plants. A large amount of soil and rock that moves downhill all at once is called a mass movement. Landslides are mass movements. Heavy rain and earthquakes often cause mass movements. Wind also erodes rock. It carries light rock particles such as dirt, sand, and clay to new places in a process called **deflation**. Deflation occurs mostly in dry areas, such as deserts, where there are no plants to keep the ground in place. Wind also erodes rock through abrasion.

14

Making a concept web can help you to memorize features of a subject. The subject goes in the middle. Things you know or need to learn about the subject go around it. Here the subject, erosion, is surrounded by facts about and pictures of things that cause erosion.

Concept Web: Erosion

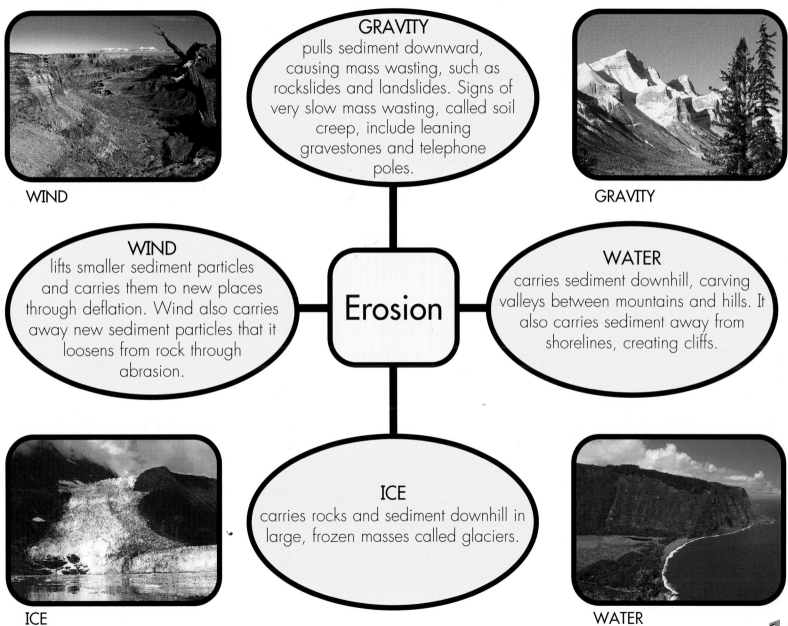

WIND

GRAVITY
pulls sediment downward, causing mass wasting, such as rockslides and landslides. Signs of very slow mass wasting, called soil creep, include leaning gravestones and telephone poles.

GRAVITY

WIND
lifts smaller sediment particles and carries them to new places through deflation. Wind also carries away new sediment particles that it loosens from rock through abrasion.

Erosion

WATER
carries sediment downhill, carving valleys between mountains and hills. It also carries sediment away from shorelines, creating cliffs.

ICE
carries rocks and sediment downhill in large, frozen masses called glaciers.

ICE

WATER

15

KWL Chart: Erosion

What I **K**now	What I **W**ant to Know	What I Have **L**earned
• Many rocks and pebbles in streams are smooth and round.	• How do rocks in streams become smooth and round?	• Water pushes larger stones and pebbles along the bottoms of streams. These rocks bump into other rocks, sediments, and each other. The bumping, a form of abrasion, weathers their surfaces evenly until they are smooth and round.
• Many houses are built on steep hillsides.	• Why don't houses built on steep hillsides slide down the hill from erosion?	• People have several ways of slowing and stopping erosion on hills. They can make the hill stronger by putting up walls or steel nets. They can make the hill less steep. They can use plants and trees to help keep the soil in place.
• Wind can blow dirt and sand into my eyes.	• How do dirt and sand get into the air?	• When wind blows fast enough, it can lift small bits of sediment, such as dirt and sand, into the air. This only happens in places with dry ground. Wind cannot lift sediment that is covered by plants or made moist by water in the ground.

Water and Erosion

Water is one of the most powerful forces of erosion. Streams, oceans, and glaciers all erode rock. Any flowing body of water is called a stream. All streams carry rocks and sediment. As rocks bump and wear away the bottom and sides of a stream, they weather more rocks and cause more erosion. Over thousands of years, streams erode enough rock to create valleys. A stream created the Grand Canyon. Streams also erode rock under ground, creating caves. Oceans erode shorelines. Waves crash against the shore, breaking down rock and moving sand on and off the beach. Ocean currents carry away sediments. Over time waves can erode entire islands. Water also freezes into huge masses of packed snow and ice. When the mass begins to move downhill, it becomes a glacier. As it moves, the glacier picks up and breaks down rocks, carving a valley into the mountainside.

This graphic organizer is called a KWL chart. A KWL chart helps you to find out what you already know about a subject, what you would like to know, and what you have learned after you have studied a subject. This KWL chart answers questions about erosion.

Deposition

Weathering and erosion break down and move rocks and land. However, they also build up land. Once sediment has been formed and moved, it settles and makes new landforms. This is called **deposition**. Sediment moved by streams is called the stream load. As a stream slows down, the heavier sediment particles in the stream load settle to the bottom. Lighter particles, such as sand and clay, travel with the stream until it reaches a lake or an ocean. When stream waters meet ocean or lake waters, the stream stops flowing. It deposits the rest of its load, forming a **delta**. Ocean currents carry away some of the sediment and deposit it on beaches. Glaciers deposit sediment, too. When a glacier melts, it deposits a mix of boulders, pebbles, sand, and clay called till. Wind also moves and deposits sediment. The stronger the wind, the larger the particles it can move.

A sequence chart shows the steps of a process that has a beginning and an end. The arrows point from step to step. Making a sequence chart can help you to remember the order of the steps in a process, and how they work together to get something done.

Sequence Chart: Formation of a Sand Dune

Wind blows sand onto a bush or a rock, starting a small sand pile.

↓

Wind deposits more sand on the pile.

↓

As wind continues to deposit sand on the pile, the pile becomes a sand dune. Dunes can be as high as 328 feet (100 m).

↓

Wind continues to blow sand off the front of the dune and deposit it on the back side, slowly changing the shape of the sand dune and moving it.

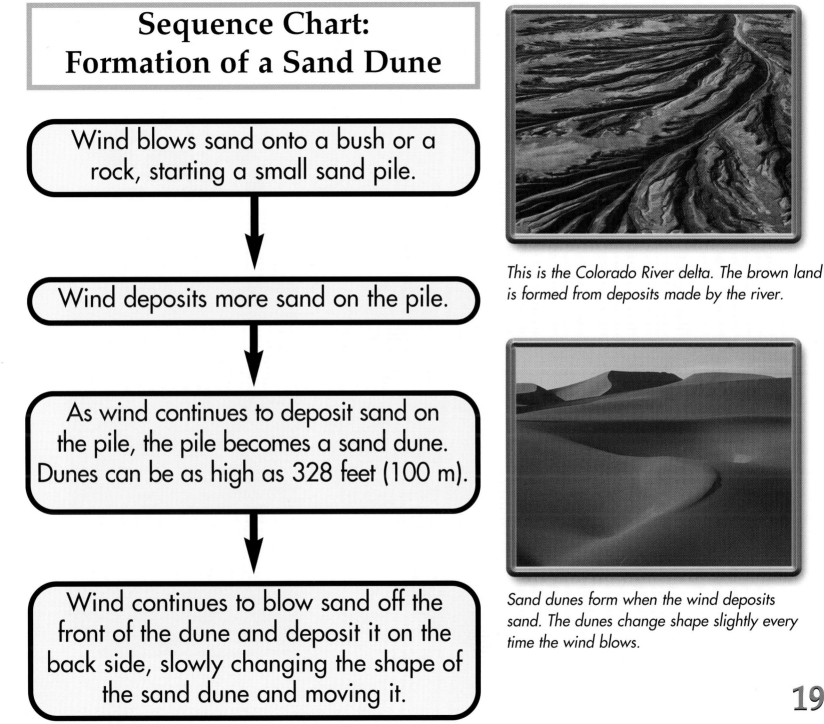

This is the Colorado River delta. The brown land is formed from deposits made by the river.

Sand dunes form when the wind deposits sand. The dunes change shape slightly every time the wind blows.

19

Pie Chart: Make-up of Soil

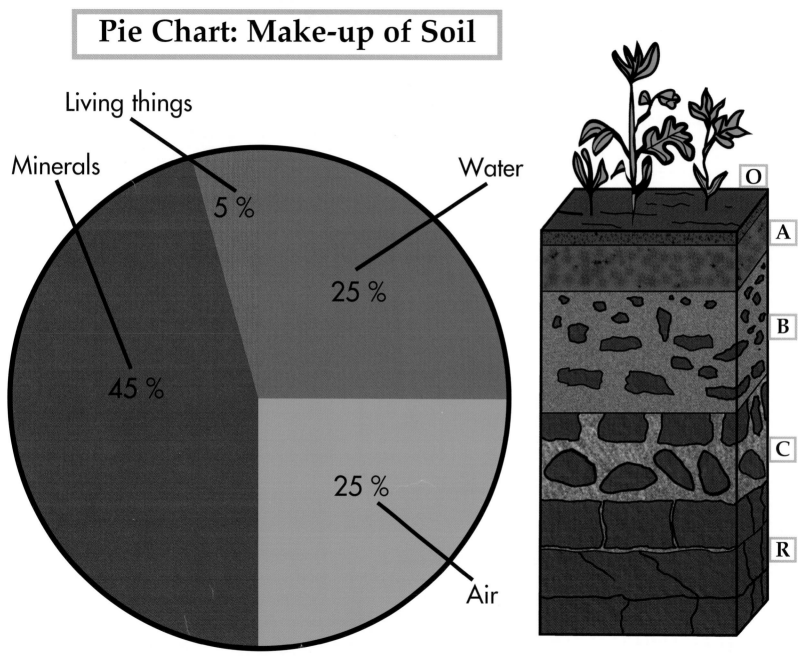

Living things

Minerals

5 %

Water

25 %

45 %

25 %

Air

O

A

B

C

R

Soil

Soil forms from weathered rock. It is a mixture of sediment and **humus**, a dark brown mass formed from the remains of dead animals and plants. Humus helps soil to stick together and hold water. Some soils form as the rock beneath them weathers. Other soils form from rock particles that have been moved by erosion. The rock that forms the soil is called the parent rock. All solid rock underneath soil is called bedrock. Sometimes the bedrock is also the parent rock. Soil also contains air and living things such as worms. Soil forms in layers called **horizons**. The top horizon, O, is mostly humus. The next horizon, A, is called **topsoil**. It is made of rock particles, minerals, and humus. Most plants grow from topsoil. The next horizon, B, is clay and larger rock particles. The next horizon, C, is partly weathered rock. The bottom horizon, R, is bedrock. Soil can take thousands of years to form.

Left: *This is a pie chart. It shows what soil is made of, and what percent, or portion, of each thing is found in an average sample of soil.* Right: *This diagram shows the layers of soil. The horizons, from top to bottom, are called O, A, B, C, and R.*

Learning About Earth

People learn about Earth through the study of rocks. By studying rocks, scientists are able to understand how and when Earth was created and how the planet has changed over billions of years. Scientists who study rocks are called geologists.

Sedimentary rock is the easiest rock to study, because it forms in layers. Each layer tells scientists something about how it was deposited and what the climate was like at that time. Fossils trapped in the layers also teach scientists about life on Earth. Fossils are the remains of plants, animals, and other living things that were buried in layers of sediment. Fossils show scientists what kinds of life existed when the rock layers formed.

As rocks form, weather, erode, and are deposited around Earth, they leave valuable clues about the planet. Rocks continue to change form today, just as they have done for billions of years.

Glossary

abrasion (uh-BRAY-zhun) The rubbing together of rock and tiny bits of rock.

acids (A-sidz) Liquids that break down matter faster than water does.

chemical (KEH-mih-kul) Matter that can be mixed with other matter to cause changes.

compound (KOM-pownd) Two or more things combined.

crystalline (KRIS-tuh-lin) Formed in a regular pattern.

deflation (dee-FLAY-shun) The movement of soil, sand, and dust by wind.

delta (DEL-tuh) A pile of earth and sand that collects at the mouth of a river.

deposition (deh-puh-ZIH-shun) The dropping of tiny bits of rock in a new place.

dissolve (dih-ZOLV) To break down.

element (EH-lih-ment) The basic matter that all things are made of.

graphic organizers (GRA-fik OR-guh-ny-zerz) Charts, graphs, and pictures that sort facts and ideas and make them clear.

gravity (GRA-vih-tee) The natural force that causes objects to move toward the center of Earth.

horizons (huh-RY-zunz) Layers of soil.

humus (HUH-mis) Dark brown matter formed from the remains of dead plants and animals.

igneous rock (IG-nee-us ROK) Rock formed by the cooling and hardening of hot, liquid rock from under Earth's surface.

magma (MAG-muh) A hot, liquid rock underneath Earth's surface.

mass wasting (MAS WAYST-ing) The downward movement of rock and soil.

metamorphic rock (meh-tuh-MOR-fik ROK) Rock changed by heat and heavy weight.

particles (PAR-tih-kulz) Very small pieces of something.

sediment (SEH-dih-ment) Gravel, sand, silt, or mud carried by wind or water.

sedimentary rock (seh-dih-MEN-teh-ree ROK) Rock formed by layers of gravel, sand, silt, or mud pressing together.

temperature (TEM-pruh-cher) How hot or cold something is.

topsoil (TOP-soyl) A layer of soil at the surface of Earth that is important to plant growth.

Index

A
abrasion, 13–14

B
bedrock, 21

D
deflation, 14
deposition, 18

F
fossils, 22

G
glacier(s), 17–18

H
horizons, 21
humus, 21

I
ice, 5, 13–14, 17
igneous rock, 9

M
magma, 9
mass movement, 14
mass wasting, 14
metamorphic rock, 9
minerals, 6, 10
mud, 5, 9

O
oxidation, 10

P
parent rock, 21
particles, 6, 13–14, 18, 21

S
sediment, 9, 14, 17–18, 21–22
sedimentary rock, 9

T
till, 18

Web Sites

Due to the changing nature of Internet links, PowerKids Press has developed an online list of Web sites related to the subject of this book. This site is updated regularly. Please use this link to access the list:
www.powerkidslinks.com/gosci/rocweaero/